The Invisible Human

Hassan Rasheed

The Invisible Human

For further information, contact:
ostaahmed@yahoo.com

ISBN
978-1-4583-2484-9

Published in the United States of America

By Lulu.com

The Invisible Human

Forward

If you look at a forest you will most likely not see the white-tailed deer that live in it. That does not mean they are not there. It only means they are invisible.

The Invisible Human

Prelude

4 billion years ago the Sun sent its rays to Earth and the Earth used this energy budget with frugality by cascading it through plants, animals, insects and microbes constantly recycling Earth's minerals in this process. But when modern humans appeared on the scene as the civilized, chaos broke loose. They discovered alternate energy sources and were not beholding to the Sun anymore. They dug deeper into the Earth scattering its minerals speeding up the advancement of entropy as if aliens had landed on Earth and implanted a foreign seed in the brains of a tribe of humans who eventually took over.

These humans appeared as bullies swaggering their way throughout planet Earth without regard to its original inhabitants in pursuit of dominance which they called their happiness. They didn't understand that their so-called pleasures were at the expense and demise of other species, the Earth itself and eventually themselves.

The intent of this expose is to convey the idea that Nature is, for a lack of a better word, the only authority.

How Nature Intended Things to Work on Earth

Earth has a heartbeat. As the Sun rises in the sky the beat starts with life springing into action moving around and playing with molecules and as the Sun sets the beat slows and life takes a break.

Earth is a most self-contained system. Hardly anything comes into it from outside, with the exception of sunlight and the occasional meteor; rarely anything leaves it with the exception of a little reflected sunlight and an occasional molecule of air floating into outer space.

Earth has cycles!

As the story goes, "An atom or a chemical compound moves through its environment, starting in one part and moving to another, only to return back to where it started with very little in terms of scatter or accumulation." There is an endless number of cycles taking place on and within the Earth right now.

The cycle that can be perceived most readily is the water cycle which begins with the Sun's warmth in the oceans, lakes, rivers and on land — as water vapor ascending through the air and forming clouds. These clouds will, in turn, drift along until the moment is right and they give in and fall in the form of rain, snow, or ice onto land to form streams, creeks, and rivers eventually emptying into lakes and oceans or else are absorbed by the land.

The water cycle takes a year or more to reach completion and there are other cycles that may take less time or much longer. For example, The Earth's crust consists of tectonic plates of rock, sand, and mud moving very slowly and colliding with one another. Science tells us that at their collision points, one plate with all its mud and sand can slip under another, producing extreme pressure and heat. This results in volcanic activity that throws molten rock from deep beneath up to the surface. The rock cools and over millions of years breaks down to mud and sand once again.

Carbon dioxide exists in the oceans, air, and underground. There is a balance of carbon dioxide between the atmosphere and the oceans, as the atmosphere and the oceans exchange carbon dioxide according to their temperatures and composition.

Scientists like to talk about these as carbon dioxide cycles. A major cycle in this regard is the photosynthesis/respiration cycle of living matter in the biosphere. For plants, photosynthesis takes carbon dioxide and water to produce Oxygen, sugars and starches upon which many if not most of the living depend on for energy. With respiration, the reverse can be observed: Oxygen and carbohydrates are combined, releasing carbon dioxide and water once again.

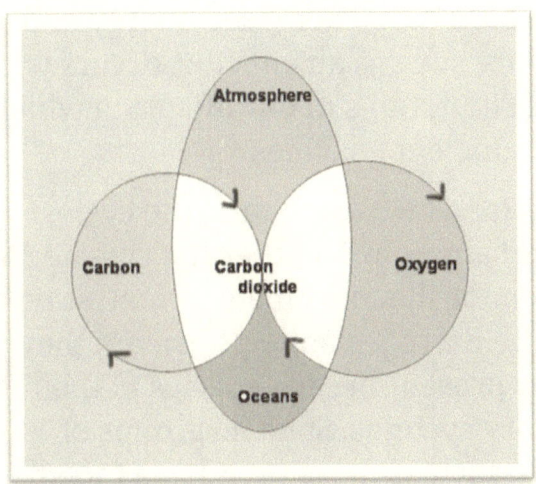

Example of the carbon dioxide Cycles

The Living Cycles

In simplest of terms, life is a chemical reaction requiring energy. With a fixed amount of incoming energy from the Sun gracing the planet's surface, lowering the amount of energy required for this chemical reaction to take place allows for a proliferation of life. If the reader asks how they might lower the amount of energy required for a chemical reaction, the answer would revolve around the availability of catalysts or enzymes. In most cases where an enzyme is required, reactions occur faster because they require less energy to activate. You can think of an enzyme as a tool like a wrench which makes it so much easier to unscrew a nut.

In summary, increased energy efficiency using tools, such as an enzyme, occurs when there are mutations in our offspring which lead to more efficient enzymes.

As far as we are aware, the cycles discussed above are neither the rule nor the exception for the movement of matter on and within the Earth's mantle. However, they are certainly present in each species in their circulatory systems. Specific groups of species—such as those found in a forest—circulate matter too. One can say that energy drives the circulatory system of forests, for example. Without it, forests would collapse because each of its links, whether one species or a piece of inanimate matter, provides a tiny link in the recycling of life-giving nutrients. This makes them available for the regeneration of the forest.

When looking out a window or strolling through a park, the reader may observe trees, birds, grass, and bees, to name just a few of the life forms present in our surroundings. When focusing on a flower merely as an object, it may appear entirely independent of the other objects around it, but in reality, it is intimately connected with all other objects in its immediate vicinity.

A clover, for example, needs soil to grow. As a mature plant, it later provides nourishment for the bison which roam the plains of the Midwest. The puma, meanwhile, later hunts out the bison for food. When it manages to bring one down, it drags what it can of the carcass to the shade of a tree to eat. Flies gather around the carcass to feed, lay eggs and try to avoid the birds that feed on them.

The above picture and explanation are obviously simplified, but they show how different species are so closely interlinked. A more accurate representation of the cyclical link between species is shown in the following graphic.

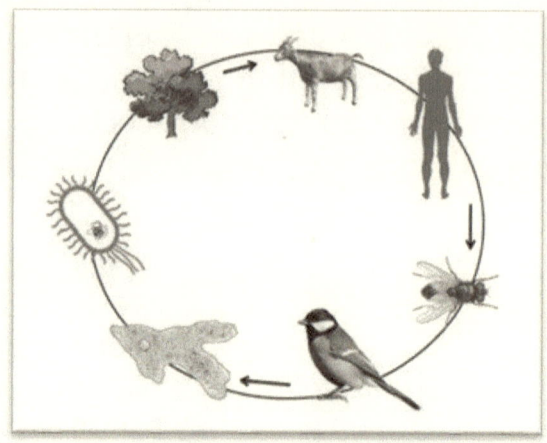

When following the biological cycle pictured above in a clockwise direction, starting with the green bush at the top left, a story can be told of how nutrients flow from one species to another. First, the bush extracts water and minerals from the soil and with the help of the Sun and atmosphere produces tender leaves which are consumed by the goat to build its muscle, maintain its growth, and hopefully also give birth.

The goat in turn is consumed by the human in the picture. The leftovers from the goat are used by the fly to lay its eggs. Some birds will feed on those flies, while the goat, human, bird, and fly leave behind fecal matter that is consumed by various microbes such as amoebas and bacteria, which return the nutrients back to the soil once again.

On any given territory, there are multitudes of life cycles, but perhaps with different species that participate in returning the atoms and chemical compounds (nutrients) back to the soil and into the atmosphere. In addition, these living cycles can often be interrelated. For instance, a living cycle that starts with a bush can have more than one grazer consuming it: they may be goats, sheep, or deer. These living cycles can take the following shape:

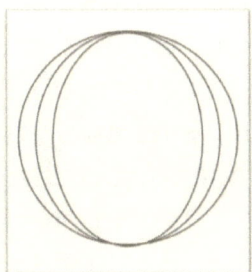

Looking at a square acre of land and trying to describe the number of living cycles within it, one may count many millions of combinations, their arrangement may look similar to the living cycles in the next diagram:

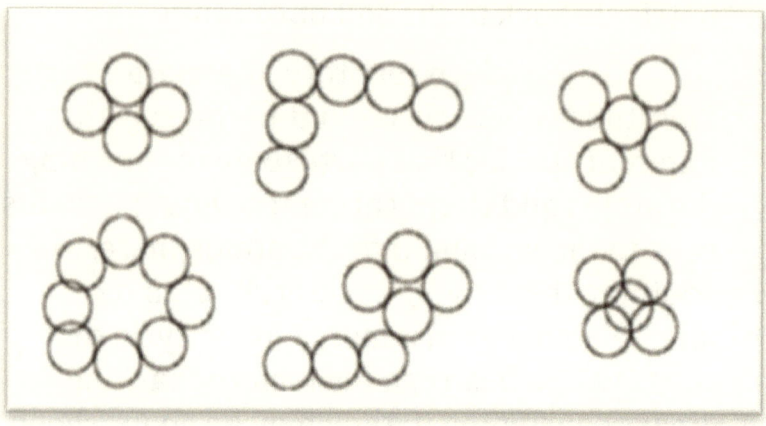

Living cycles are not limited to food, either. They can be observed in relation to nest building, territory, and so on. The biological cycle graphic below shows the cycle for a bird building a nest in a tree. When the eggs hatch and the fledglings fly away, the nest may fall to the ground where insects break it down and pass on the resulting matter to amoebas and bacteria which will return the nutrients back to the soil to be used by the bush.

This is a key point to introduce in any discussion the concept of an ecosystem. A forest is an example of an ecosystem: Like any ecosystem, it contains mega cycles of circular chains involving both the air, soil, water, plants, and animals. A forest is not the only ecosystem. A desert such as the Sahara Desert is also considered an ecosystem.

In summary, the chain of life is circular, meaning that there is no start or end, no top or bottom link, nor is there any intelligent link or any particularly unintelligent one. The chain is nevertheless very complex. It contains many circular side chains. The chain is like a solid sphere where no one circular chain can be distinguished from another.

The Mature Species

When looking at the relationship between a lion and its prey—the gazelle for example—

the African lioness eats from the herd of gazelle to satisfy her hunger and the hunger of her pride. She hunts in a way that always leaves enough gazelle for her to pursue the next day and the next year. If she and her cohorts finished off the whole herd of gazelle, then her cubs will have nothing to eat the next day or year and they will die off, resulting in their extinction.

One key factor that gives stability to this living relationship is "laziness". In the wild, it has been observed that lions sleep for 21 to 22 hours a day and go hunting for gazelle or other prey only when their bellies rumble. To conserve energy, lions pursue only the slowest individuals in the gazelle herd. If an element of laziness was not present in this pride of lions, one can imagine the lions hunting all day and bringing down as many gazelles as they can, which would ultimately break the living relationship between predator and prey.

Other key factors to consider here are the territoriality of lions and the increasing speed of gazelles that protect themselves against overhunting.

In summary, living relationships are stable over time largely due to the genetics of their participants which have mutated into a form that fits the environment. Unstable relationships in nature simply do not exist.

How Nature Intended Humans to Live

Humans continued the 4-billion-year tradition of forming flesh and blood out of the rays of the Sun and a few pounds of Earthly minerals which they then returned back to the soil when life ends. Humans were successful in upholding their place in this marvelous cyclical endeavor for thousands of years as predator, prey or any of the many relationships that sustained nature's enterprise.

By civilized standards, nature's marvelous undertaking of evolution was an un-intelligent venture. You see, nature operates on a hit and miss bases as opposed to the civilized way of learning and planning. As a result, maybe 1% of species had populations that were stable enough to prosper while the 99% failed to do so.

This stability usually depended on other longstanding species in the environment that helped it maintain steadiness through a death rate that was equal to its birth rate. By becoming a consumer, it hunted or gathered neighboring species for sustenance and by becoming the consumed, its population found a haven from the scourge of over population.

Since the rays of the Sun was the only source of energy in pre-civilized times, stability was also swayed by efficient energy use which in turn influenced species body size, shape and behavior as well as population size and density. It did not matter to nature whether a species spent most of its time sleeping like the African lions or spending most of their daylight hours eating such as the gazelle so long as the energy the lion or gazelle were intaking fit their needs to maintain species stability and survivability.

Now, we come to the early humans of whom we know little of how they lived. We have glimpses from current indigenous human tribes, prehistoric cave paintings and our closest relatives the apes. What information we can glean from these sources is that early humans probably lived in scattered nomadic tribes consisting of approximately 15 to 100 individuals each with overall densities of approximately one square kilo meter per individual.

Agriculture had not been invented yet and so they depended on the environment to provide them sustenance with seasonal foods such as roots, grains, berries and fruits with an occasional piece of meat. They lived in warmer parts of the world as opposed to colder iced regions. They probably made temporary shelters from local natural materials leaving them behind when they moved. They made simple tools to help them acquire the resources they needed. These tools included spears, nets and some tribes had poisonous blow darts.

They had a leisurely life style spending approximately 2 to 3 hours a day hunting and gathering while spending most of their time socializing, being creative around food preparation, eating and celebrating into the night with theatrical song and dance. This type of routine probably kept them from over hunting and becoming a burden on the environment they inhabited.

From what we could tell, they had a close connection with their environment. For example, today's pigmies of the Congo River Basin consider the forest they live in as their protector often calling it their mother or father. They are so integrated and in balance with their forest environment that a stranger looking at the forest from the outside would not guess there are people living there. But the forest pigmies are there. They are just invisible.

The Invisible Human

How Humans Ended Up Living

All living individuals whether human or not have this idea that they are the most important entity in the world. This impression was the impetus behind self-preservation, the conservation of the species and on a grander scale the protection of the whole complex chain of life.

But a few thousand years ago this self-preservation drive was influenced by a group of changes in the key of life of a human population that caused the emergence of the civilized who diverged from the cyclical traditions and created a world of their own. They developed concepts such are those of gods that bestowed upon them the right to own the Earth and all it contained from the living to the mineral. They delved into technologies based on physics, chemistry and biology to develop industrial economies that made the appearance of this superiority self-evident.

Instead of seeking a simple lifestyle that pursued humble pleasures, they created political systems to govern the distribution of wealth to their populations. But managing wealth was never an evolved task attributable to humans nor were the concepts as capitalism or free trade. Unfortunately, those practicing capitalism prayed to the gods of profit every second of every day. In a zero-sum world, profits had to come from somewhere. This somewhere was the body of the Earth and its inhabitants resulting in the robbery of the Planet's dignity.

For example, the civilized patted themselves on their backs when they slaughtered what they called the beast until the beast became extinct. They patted themselves on their backs when they claimed victory in wars over their own kind. They cleared whole forests using its wood to build monuments to themselves and again patting each other on their backs for a job well done. They did not let the forests grow back either but sequestered the land for artificial food growing fields to feed their growing populations.

And these lands required large quantities of water for irrigation. Rivers were stopped from running free to the demise of the aquatic species that depended on their fresh flows. They dammed the salmon and other migratory species from reaching their spawning ponds and they dammed the young from reaching the open ocean waters to mature and repeat their life cycles once again.

They did not leave things alone at that either. They broadcast herbicides, pesticides and other chemicals over the lands in order to keep up the production levels they needed to house, clothe and feed their growing populations and called it progress. It did not matter to them that other species suffered or perished and they remained unaware they were harming themselves in this tragic process.

In addition, the civilized decided to gouge out and upend the land in search of minerals they deemed necessary to support an unrealistic lifestyle. An existence that was in constant need of widgets that failed to satiate the hunger for more.

They did not rest there but continued to explore deeper in the ground and found a crude source of energy that allowed them to compound the damage they were already committed. They felled more trees, took the lands to plant additional crops, dug additional gigantic cavities in search of extra minerals. They pumped still more crude into their economies to the point that they, the civilized, with their domesticated animals and pets, constituted 96% of the biomass of all mammals on the planet.

All the above activities resulted in further injury to Earth as the top soil eroded away, river waters became poisoned and depleted, glaciers melted and the air became polluted causing huge flocks of birds to fall dead out of the sky. The damage continued into the oceans that became the dumping bodies for garbage; where massive trolling depleted fisheries and the danger of many species going extinct was becoming real.

They felt they had the need to eliminate or postpone death from their lives. This caused the equation of a sound population balance with the environment to unravel. They were blinded by emotions and the inability to perceive the fact that death was the only way to insure the start of a fresh beginning by rebirth on a finite and ever-changing planet.

The disrespect did not stop there but now the civilized are aiming to conquer the heavens in search of minerals exhausted here on Earth. They also aim to construct whole industries in outer space in the name of saving the Earth from disaster and patting themselves on their backs for coming up with this grand scientific driven decision.

Because of the explosion in its population counts and high-density rates the civilized are overwhelmed by 81 communicable diseases. A reduction in the concentration of people is needed from 100 individuals per square kilo meter to 1. But then where would the excess go?

It all boils down to faith and the appreciation of nature over our desires for capitalism's promise of happiness and freedom. Unfortunately, capitalism isn't freedom. Its basic premise is the accumulation of wealth through increased profit. So, capitalism must enslave, kill, cut down trees, dig huge hollows in the ground, acquire other species habitats or pollute the land, sky and water in order to provide a profit and a short-lived mirage of freedom.

In reality, capitalist corporations are just providing for the demand the civilized put on them. What was needed early on was a grass roots effort to change our lifestyles to one that appreciated the free gifts given to us by nature. If they had done so these corporations would have disappeared and so would their capitalist drive and false promises.

Today, indigenous tribes like those of Amazon River Basin are the salvation of the planet Earth. They live without damaging their environments. They have a nature based ecological footprint. It is criminal what is being done to them and their habitats in the name of civilized freedoms.

Unfortunately, it is too late to make any efforts to change now. The civilized have painted themselves into a corner with no way out. They are accustomed to their power, wealth and comfort to see any avenue of "regression" to a simpler lifestyle supported by nature. It is no secret that nature is the civilized arch enemy and they have set their delusional hopes on science to find a way to defeat it.

What Went Wrong?

Nature is complex. Humans are, in fact, incapable of grasping its full intricacy. The reality is that humans are but one link in the chain that is of equal value to any other link in the chain of life. The civilized must rid themselves of the notion they are more intelligent or somehow superior to any other life form. The civilized must learn humility and respect for the order and boundaries of nature.

With precious Earth on the cusp of catastrophe, we need to come to the realization that we, the culprits, have somehow evolved to be this way. And I can't help but feel a deep sense of guilt and wonder as to whether the rest of the "civilized" people of this planet must feel just the same. After all, are we not, Sapiens, the "wise" ones who should learn to change our habits in order to accommodate Earth's way.

I have come to understand the human brain and its two most basic limitations: the first being a physical limitation in the quantity and quality of information it is able to store; the second is a result of the limitations of its sensory organs in perceiving reality. The brain is limited in its ability to record all of this information.

With respect to the sensory organs, the eye can only perceive a very limited band of light, known as the "visible spectrum." The same sort of limitations exists in relation to time, sound, touch, smell, and hearing. Therefore, the brain is not able to detect the full scope of reality surrounding it. For example, if the brain can store 50% of the information that is out there in the real world and our sensory organs can process only 50% of what reaches them then the total product of information a person has is 25% of reality.

Although our brain's memory is reduced in information, the brain is able to manipulate information and even synthesize it. This is perhaps best demonstrated in how we solve problems. For example, by bringing together two different memories and producing a third that may solve some need. Unfortunately, since memories are limited in information and the process of linking two of them follows no natural law, the end result is an abstracted and artificial solution that will not necessarily prove to work in the real world.

We should now realize that the structures of the brain and those of reality are quite different things. Let us imagine that an understanding of the artificial world created by the human brain comes in the form of square blocks, with natural counterparts coming in the form of a ball. We might well imagine the blocks of our imagining to tumble and fall if built on reality. When building a square structure on a round reality, it immediately becomes unstable.

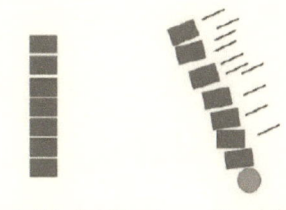

Stable Artificial Stack   **Unstable Artificial Stack When Built on the Real World**

According to such logic, anything humans can conceive of or build is ultimately doomed to be short-lived. We should rather try our best to look to the structures built by nature which inevitably proves more durable than anything we could have built. In whatever we do, we should let nature set the path because nature is the final authority on what is right and suitable.

We should have been wise and considered ourselves part and not apart from Earth's living cycles, the web of life and its ecosystems. Notions that "humans are at the top of the food chain" and "humans have a superior intellect" are vocalizations of a drunken ego inebriated by technology. We are but a link, no better or worse than any other, be it a clump of mud or another species in Earth's plethora of life forms. The interference of our machineries has derailed our egos with unrealistic ideas, leading us astray and blinding us to the fact the more we took to synthesize the more we took from the real world leaving less of it to flourish like it always had.

Can Humans Return to Nature?

Current evidence indicate we are on the cusp of a global biotic catastrophe possibly turning back the clock to climate conditions that existed 60 million years ago where global temperatures were higher as well as having a different atmospheric chemistry. There is no question human activity is driving this end result by exhuming ancient remains and resurrecting the conditions in existence back then.

The Amazon is burning because we aren't good stewards of the land. The seas and oceans are becoming depleted of life because we are not good stewards of them. The air we breathe is becoming polluted because we are not good stewards of it.

We were never meant to be stewards of anything but to follow nature's example. We were meant to live as the dwindling indigenous tribes around the world live. We needed to relax and live a life guided by nature. We needed to return to our original Sun god for sustenance, happiness and rebirth to prolong our days on this beautiful planet.

Nature does not speak like we do but communicates by chemical reactions that follow the laws of physics. Humans do not have to carry the burden of guilt as many do for the upcoming biotic disaster. After all we are also reacting chemicals obeying the natural laws of physics. So, sit back and enjoy the ride with art, music, song and dance for we are on a rudderless ship and mother nature holds the secret to where we are going.

Do you think we will survive?

Hassan Rasheed

The End

The Invisible Human